James Andrew

The Harveian Oration

Delivered before the Royal College of Physicians, October 18th, 1890

James Andrew

The Harveian Oration
Delivered before the Royal College of Physicians, October 18th, 1890

ISBN/EAN: 9783337039165

Printed in Europe, USA, Canada, Australia, Japan

Cover: Foto ©berggeist007 / pixelio.de

More available books at **www.hansebooks.com**

THE

HARVEIAN ORATION,

DELIVERED BEFORE

THE ROYAL COLLEGE OF PHYSICIANS

OCTOBER 18TH, 1890.

BY

JAMES ANDREW, M.D.Oxon. F.R.C.P.

HONORARY FELLOW (FORMERLY FELLOW) OF WADHAM COLLEGE; PHYSICIAN
TO ST. BARTHOLOMEW'S HOSPITAL.

LONDON:
PRINTED BY
ADLARD & SON, BARTHOLOMEW CLOSE.
1891.

DEDICATED

WITH PERMISSION

TO

SIR ANDREW CLARK, Bart. LL.D. M.D. F.R.S.
PRESIDENT OF THE ROYAL COLLEGE OF PHYSICIANS,

BY

HIS GRATEFUL FRIEND

THE AUTHOR.

THE HARVEIAN ORATION.

THE pleasure, Sir, with which I learnt from your
lips that I had this year to discharge the time-
honoured functions of "Harveian Orator" was
largely alloyed with dismay at the difficult nature
of the task before me. My predecessors have been
so numerous (this commemoration of benefactors is
held to-day for the 172nd time), they have also been
so well equipped for the work by literary skill, by
training and by knowledge, that it might well be
thought that nothing could be easier, or more certain to be successful, than to follow faithfully in
their steps. But, unfortunately, the very number
of my predecessors and their individual excellence
render this course impossible : for, it is one thing
to imitate with more or less success the style and
method of a great writer or speaker, it is another
and a very different matter to reproduce his actual
words and facts, to publish, so to speak, a new
edition of his work with no real change except in
the title page. Now, Harvey's shield has been
burnished so often and with such sharp-sighted
devotion that no spot or stain remains upon it,—

his claim to be the discoverer of the circulation of
the blood has been fully established, his writings
have been found to be pregnant with the ideas and
discoveries of later generations;—nay, more, the
man himself has been proved to be not less than his
work. His great discovery was no chance find, no
mere lucky hit, but the natural outcome of his genius
and of his time. How then can the belated Harveian
Orator of the present day hope to add even one
fresh pebble to the cairn which the love and the
labour of generations have raised over the "im-
mortal" Harvey.

On beginning my quest for something new it
soon became clear that, of the two fields more especially open to me, Sir James Paget had practically
exhausted that of St. Bartholomew's Hospital in his
'Records of Wm. Harvey in extracts from the
journals of St. Bartholomew's Hospital; with notes
by James Paget,' published in 1846, but I still
hoped to discover, by the help of learned friends,
some hitherto unnoticed records of Harvey's life in
Oxford. Nor, indeed, is it impossible that such do
exist although I have failed to unearth them. The
sole result of my search, and this I owe to the kindness of the present Warden of Merton College, is
the following brief report, in the Register of Merton
College, of Harvey's speech to the Fellows on April
11th, 1645, two days after his admission to the
Wardenship.

"Dominus Custos, Convocatis in Altâ Aulâ Sociis,
hæc verba ad illos fecit. Forsitan decessores Custo-
diam Collegii ambiisse, ut exinde sese locupletarent,

se verò longè alio animo—nimirum ut Collegio lucro
et emolumento potius foret ; simulque socios, ut con-
cordiam amicitiamque inter se colerent, sedulò sol-
liciteque hortatus est." This fixes the exact date of
Harvey's admission to the Wardenship, viz. on
April 9th, 1645—O. S. : and there is another reason
for my dwelling on what may seem to be a
matter of but small importance. On the strength
of this short entry in the College register it has
been stated that Harvey delivered a somewhat
" Pharisaical " speech to his Fellows. Now, it we
bear in mind the violent passions aroused on both
sides during the Civil War, nay even if we judge his
words by the modern standard of parliamentary
language as established during this last session,
Harvey can scarcely be held to have indulged in ex-
cessive abuse of his predecessors, including Sir
Nathaniel Brent, or to have said too much in praise
of himself. The exhortations to concord and good
fellowship come with double weight from a man
who cared first for learning and science and then
made provision also for a " general feast once every
year," and, "at every meeting once a month, for a
small collation, as the President shall think fit, for
the entertainment of such as come."

By a very natural feeling on the part of the
College the lecture and feast, which Harvey guile-
lessly founded in commemoration of the benefactors
of the College, have come to be devoted almost
entirely to the commemoration of the most illustrious
benefactor of them all, himself.

The value of Harvey's great discovery can scarcely

be exaggerated, either in itself, or as an illustration of scientific method and work, or in its effect on the course and development of physiological studies. There is no need to dwell on points such as these before this audience. We are all of one mind in the matter.

But, besides these great things, Harvey's writings contain also many things, of less importance indeed, but yet well worthy of our consideration, *e. g.* we may find in them no doubtful indications of the wise and prudent advice which he would have given us for the abatement of some at least of the evils under which our Profession labours at the present day. Let me give one or two illustrations of my meaning, taken from the first sixteen chapters of the treatise on the motion of the heart and blood.*

In the first rank of the ills which beset the profession I do not hesitate to place the enormous amount of so-called medical literature. Were it all good there would still be quite enough of it. But could we get rid of the vain repetitions of all those books which contain little or no original work, and are too often the damaged seed, not the fruit, of practice and of a ripe experience—could we get rid of all the clinical observations with the idle theories based upon them, which from their imperfections are simply misleading—could this be done, the residue would no doubt be small, might, indeed, bear much the same proportion to the original mass as

* The works of William Harvey, M.D.Eng., translated from the Latin, with a life of the author, by Robert Willis, M.D.Eng., Sydenham Society, London, 1847.

fine gold does to the ore from which it has been
extracted, but it would be far more valuable than
anything which medical literature in its present
form can offer to us. There is one publication
which I regard with especial horror, the "invalu-
able" 'Index Medicus.' I find that the volume for
1889, the last completed but not the greatest of
eleven volumes, contains the names of not less than
13,870 contributors. Now the index of the 'Index'
fully bears out the supposition that each pair of contributors represents not less than three communica-
tions, and thus we have a total of not less than
20,000 contributions to medical science for last year.
I have much pleasure in admitting that many of
them are short, but then it must not be overlooked
that very many of them are substantial treatises in
two volumes or more. That one volume of the
'Index' alone contains sufficient material for a small
library of monographs and systematic treatises,
which a very moderate amount of mechanical in-
dustry could easily put together. I venture to
think that no small part of medical literature would
never have been written had its authors given heed
to Harvey's words at the end of the Dedication of
his 'Treatise on the Motion of the Heart and Blood.'
"My dear colleagues," he writes, "I had no purpose to swell this treatise into a large volume by
quoting the names and writings of anatomists, or to
make a parade of the strength of my memory, the
extent of my reading, and the amount of my pains;
because I profess both to learn and to teach ana-
tomy, not from books but from dissections; not from

the positions of philosophers but from the fabric of nature—I avow myself the partisan of truth alone; and I can indeed say that I have used all my endeavours, bestowed all my pains, in an attempt to produce something that should be agreeable to the good, profitable to the learned, and useful to letters."

Again, a very common reproach cast upon our profession is that we pretend to be, and are not, scientific; that, to quote an old sarcasm, our occupation consists in putting drugs, of which we know nothing, into bodies of which we know less. There is just sufficient truth in such criticisms to make them somewhat unpleasant to those for whose good they are no doubt intended, but they derive the keenness of their sting from our own shortcomings and mistakes. We claim too much and do too little for medicine; we forget that our duty as medical men is twofold, to practise an art, and to study and advance a science; that we are bound to make the best use of the knowledge within our reach, and, if possible to add to it in the using; but this last is not our chief duty We dare not sacrifice the interests, the life of one single patient to the advancement of science. If our conscience were to cease to forbid us to do this, then the sooner modern science deprives us once for all of such a treacherous guide the better for the world will it be. A man may be a first-rate practitioner and yet have no title whatever to be ranked amongst scientific physiologists, as a sailor may be a first-rate navigator without being in any proper sense of the words a scientific astronomer or mathematician or physicist. Yet in

spite of this admission, the physiologist ought to be
the last to taunt us with ignorance and empiricism,
for his own science would be far more imperfect
than it now is were it stripped of all that it owes to
the results of medical and surgical practice.

One of the best, if not the best, definitions of
" Medicine " is that which describes it as "applied
physiology." If we fail to attain to that ideal, if
we are compelled daily to act upon probabilities in
place of scientific certainties, then the blame must
rest at least as much on the physiologist as on our-
selves. He fails to supply us with the knowledge
which we require, and which he alone can give us.
Medicine is thus but one of many instances of an art
stunted by the insufficient development of the
science with which it is connected, and on which it
rests. I do not mean insufficient in a general sense,
but insufficient for certain special purposes. Permit
me to anticipate a very possible criticism upon what
I have just said. It may be thought that my words
are an unfair attack upon physiology, and that,
safe from immediate contradiction, I have sought to
defend medicine by abusing the very science to
which, of all sciences, we are most deeply indebted.
Now, if I have indeed done this thing I have been
guilty of the basest ingratitude, I have ignored the
splendid services which physiology has rendered to
medicine, services which we believe to be but the
earnest of others yet more splendid in the near
future. If physiology is, as it seems to me, not yet
equal to all the requirements of medicine, that is
certainly not due to the indolence or weakness of

those who cultivate it, but to the inherent difficulties
of their task. They have done all and more than
all which we could have looked for. But at the sight
of sickness which we cannot heal we grow impatient,
unreasonable, and long for knowledge not yet within
the reach of man. We even refuse to recognise the
wisdom of those who decline to attempt to solve by
scientific means problems not yet ripe for such
solution.

However, although the object matter with which
we deal lends itself unwillingly to strict scientific
method, although, from its very nature, we are
unable to submit it to direct experiment, except in
most imperfect fashion, it may still be studied and
observed in a scientific spirit. Every case of disease
may be, ought to be, looked upon as an experiment
in practical physiology, an experiment carried out
with transcendent skill by nature herself, but which
she leaves it to us to observe, to register, and to
interpret.

Now, in this matter of scientific observation it
cannot be denied that we fall very far short of what
we might accomplish. It will do our profession no
good to dwell upon the legitimate excuses which may
in part explain our failure ; rather let us acknow-
ledge that failure, and use our utmost efforts to
remove it. Nor are leaders wanting who can teach
us how this may best be done. We need not look
beyond the circuit of our College to find men who
can, and do, carry on their daily work in this scientific spirit, approving themselves herein worthy
followers of the example of Harvey.

Surely we cannot regard Harvey the physician as
less scientific than Harvey the anatomist and physiologist, when we find him showing from "certain
familiar reasonings that the circulation is matter
both of convenience and necessity." In his hands,
the physical signs of an aneurysm, the effect of
extreme cold, the phenomena which attend "con-
tagions, poisoned wounds, the bites of serpents and
rabid animals, lues venerea, and the like," all these
are made to supply probable and cogent arguments
for the truth of the doctrine of the circulation of the
blood.

Were our clinical knowledge as carefully accurate in
its statement of facts, were it always used as soberly
and to as good purpose as Harvey's was, then there
would be, there could be, no outcry raised against
us on the score of credulity or of want of scientific
method. When he had arrived at a knowledge of
the circulation of the blood by means of anatomical
researches and vivisections, Harvey at once applied
this knowledge to the explanation of clinical phenomena up to that time inexplicable. He then used
the fact that this explanation was a simple and
adequate one as a new and strong argument for the
truth of his discovery. That this intimate connection, this solidarity, between physiology and medicine is no longer recognised so fully as it once was, is
much to be regretted, for it is injurious to both lines
of study, and has arisen, at least in part, from faults
on both sides. But this view must not be pushed
too far. We must not lose sight of the fact that
the relationship between physiology and medicine

has in many ways greatly changed during the last
250 years, and that such change is a necessary consequence of the progress made by physiology.

In Harvey's time, in any classification of the
sciences, physiology might have been regarded as a
department of medicine; I need scarcely say that
now the reverse would be the case. The two deal,
it is true, with the same object-matter, for there
is no physiological fact or law which is without some
bearing upon medicine; and, again, every medical
fact or generalisation, even those most purely empi-
rical, is more or less important to physiology. But
their aims and methods are so different that they
tend inevitably to become more and more sharply
differentiated from each other.

The goal of physiology is truth, *i. e.* perfectly
trustworthy knowledge of a certain class of facts
and laws; and this independently of any use, good
or bad, to which that knowledge may be put. The
goal of medicine is power, *i. e.* ability to manipulate
certain natural forces in such fashion as to produce
certain effects. No doubt theoretically the two
ends coincide, and we may hope that in some remote
future they will do so, in reality and perfectly. For
the present we must be content with having, in one
direction, much knowledge which confers little or
no power, and, on another side, very imperfect
knowledge which yet brings with it very great
power, too often ill-directed. Again, their methods
are different. Physiology by slow degrees has come
to rely more and more on purely scientific modes
and instruments of research, and to apply them by

preference to matters which can be brought to the test of direct experiment. Medicine, on the other hand, has no choice but to remain, so far as it has a scientific side, a science of observation, for anything like effective investigation of the matters with which it deals by direct experiment is impossible. As Physiology slowly reduces to order the apparently hopeless confusion of so-called vital actions, the easiest questions are attacked and answered first, and thus those which have to be faced later in their turn are more and more difficult, more and more refractory to scientific analysis.

Now these more difficult questions are often of vital importance to medicine, and in them lie dormant vast possibilities of increased knowledge of the nature of disease, of increased power over it. And yet from the great difficulty of subjecting them to experiment physiology may seem for a time to fail us, and the task of employing physiological results to explain clinical facts, or to form the basis of rational treatment, becomes harder than ever.

This brief sketch of the relations between physiology and medicine, and of the change which has gradually taken place in these relations, a change leading in appearance and for a time to a great widening of the interval which separates two sister sciences, is, I well know, very imperfect and full of contentious matter, but it arose naturally and took its colour from my own experience, especially from that of the last few months. Not the least important part of the charge of the Harveian Orator is that which directs him to exhort the Fellows and Members of the College to "search and study out the secrets of nature by way of experiment." It is a little embarrassing for the scholar to find himself called upon to instruct his teachers, but I think that I cannot employ the time at our disposal better or in a manner more agreeable to the declared intention of the great founder of this lecture than by laying some recent experience of my own before you.

I do this with much diffidence and a full consciousness of my own inability to deal satisfactorily with the weighty matters involved in it, and would ask you not to overlook my shortcomings but to correct them.

In the course of last winter it was my fortune, whether good or bad I know not, to meet with several severe cases of hæmoptysis, two or three of which, in spite of my best efforts, ended fatally. About the same time I rashly undertook to give a lecture on the treatment of hæmoptysis at the City of London Hospital for Diseases of the Chest. My attempt to instruct others was repaid by the highest reward which can fall to the lot of a would-be teacher, the conviction of my own ignorance. I found, not for the first time, what many of you must also have found, that the treatment of hæmoptysis is eminently unsatisfactory. In fact that, beyond a few general measures and simple remedies, all tending to reduce blood-pressure in the vascular system as a whole, it was very doubtful whether our present supposed knowledge enabled us to do any good at all to our patient. I was reduced to a state of therapeutic despair and even suspected that of the drugs commonly employed some might be actually injurious. The same feelings, the same sense of helplessness, had often passed over me before, but without any good result. This time the stimulus, perhaps of St. Luke's day, was stronger and I tried to look for a remedy. The following enumeration of remedies which, at different times, and on more or less satisfactory grounds, have been supposed to be of value in the treatment of hæmoptysis, is taken from a well-known standard work on Pharmacology, Therapeutics and Materia Medica. Long as it is a less critical author might have made large additions to it : neither have I any fault to find with the place which it, and similar lists, occupy in therapeutical writings, they are a necessary part of the work, and in this case the writer has been careful to add to its value by drawing special attention to those means on which he believes that the most reliance may be placed. It is noteworthy that he does this in three cases only, viz. gallic acid, hamamelis and lead acetate.

Hæmoptysis.

Acetic acid
Aconite
Alum
Ammonium chloride
Arnica
Astringent inhalations
Barium chloride
Chlorodyne
Chloroform. To outside of
   chest

Copaiba
Copper sulphate
Digitalis
Dry cups. To chest
Ergot and ergotinin
Ferric acetate
Ferri persulphas
Gallic acid. Very useful
Hamamelis. Very useful
Hot-water bag. To spine

Ice
Ipecacuanha
Iron.   And absolute rest
Lead acetate.   Very useful
Matico
Morphine
Opium
Phosphoric acid
Potassium bromide
Potassium chlorate
Potassium nitrate

Pyrogallic acid
Silver oxide
Sodium chloride.   In drachm doses
Subsulphate of iron
Sulphuric acid
Tannin
Tr. Laricis
Turpentine
Veratrum viride

Now, on looking through a long list like this, one's first and last thought is that it gives us a very good illustration of the truth of the old axiom that when many drugs are supposed each to cure one and the same disease, we may safely hold that few, if any of them, have the least influence over it. Some of these before us have been proposed on chemical, some on nervous or vasculo-nervous theories, and the claims of a few have been supported by direct physiological experiment, but all in the last resort profess to have had their worth determined by clinical observation of the effects which follow upon their employment.

In a matter of this kind, however, clinical observation has to be received with great distrust and must be carefully sifted. It is not sufficient to quote a long list of cases in a large percentage of which complete recovery followed the exhibition of certain remedies. With this we want a control-list of the issues of cases which have run their course without any treatment whatever, and so far as haemoptysis, or indeed any other disease, is concerned, no such control-list exists. Experience in large out-patient hospital practice for many years, taught me that cases of hæmoptysis, of all degrees of severity, are of very frequent occurrence, in which recovery takes place without any medical interference, and that too under very unfavourable conditions of life. And further, that this *vis medicatrix naturæ*, this self-help, shows its power most clearly in what may seem to be the worst cases; *i. e.* a patient has a sudden rapid profuse hæmoptysis, he remains at home, and the bleeding does not return, or only after a long interval to be reckoned by weeks, months, or even years; or, again, so soon as, or even before the hæmorrhage has ceased, he is seen by a doctor who injects ergotinin subcutaneously, and believes, pardonably enough, that the injection has stopped the bleeding. But in each case the syncope, following the sudden loss of even a comparatively small quantity of blood, has given sufficient pause to the circulation to favour the formation of a clot in the ruptured vessel itself, or in the bronchus or vomica with which it communicates.

We cannot judge of the effect of remedies without some knowledge of the natural history of the diseases in which they are employed. Perhaps in the case of hæmoptysis, the same mistake, viz. that of attributing the natural termination of a pathological process to the effect of remedies, is repeated, which up to a few years ago was made in that of pneumonia, another pulmonary affection, and of the specific fevers. Clinical observation has so far failed to give anything like scientific proof of the supposed influence of certain drugs upon hæmoptysis, *i. e.* upon the pulmonary circulation. After all it is not very surprising that clinical observation has failed to give definite results in this matter. The difficulties in the way of investigating the relative effect of drugs by any means at the command of the medical observer, may well be in many cases insuperable. But it is disappointing to find that the drugs which have been proved by direct physiological experiment to be at any rate likely, on the theories of those who propose them, to be of service in hæmoptysis, do not come more distinctly to the front, do not establish a stronger claim to our confidence than the rest. It almost seems as if in this case we could not expect to receive any help from physiology. Happily this is not so; physiology is at last overcoming the great difficulties in the way of accurate determination of variations in the pulmonary circulation and of their causes; and the results already obtained, imperfect though they be, are of sufficient interest and importance to justify me in bringing them before you to-day.

The relations between the systemic and the pulmonary circulations in respect of blood-pressure and other conditions form a large and, it may be, a very fertile province of vascular physiology which has been almost entirely ignored by medicine and treated by physiology so slightly that it has yielded but few, if any, results of value to medicine. I do not know any treatise on systematic medicine, any monograph on diseases of the lungs which contains even a hint that the two systems may be to some extent independent of each other, and may react in different ways to the same stimulus and in the same way to different stimuli.*

Throughout medicine assumes that indications for the treatment of pulmonary diseases may be taken from the systemic pulse, as directly and with as much confidence as in the case of erysipelas or of enteritis. The pulmonary circulation is treated as if it were simply the termination of the systemic venous trunks into which a small additional pump, the right ventricle with its valvular arrangements, has been introduced with the effect of preventing undue stagnation of the blood current. And this view does contain a half truth ; *e. g.* in many cases of long standing and extreme mitral incompetence we can trace, after death, everywhere as we pass backward against the blood-stream from the left auricle to the large systemic veins, conclusive evidence of the existence during life of undue vascular tension. In the left auricle, pulmonary veins, pulmonary artery, the large systemic veins, in the hepatic veins, we find thickening and opacity of the inner coat of the vessels, in fact atheroma ; and this too in cases where the systemic arteries, its most frequent seat, may be all but free from it. Throughout this tract it would seem that the same condition of undue pressure has prevailed, and that, if any forces exist tending to produce inequalities of pressure in different regions, they must be very feeble ones. Of course this is an extreme, though not infrequent case, and cannot be regarded as throwing much

* See note at the end of the Oration.

useful light upon the state of things in health or even in less severe cases of disease.

Anyhow, medicine, if it recognises any mechanism in the pulmonary circuit analogous to the vaso-motor arrangements on the systemic side, has not thought it worth while, or has not been able to make any practical use of such knowledge.

Now physiology has not done quite so much as we might wish, but it has certainly accomplished more than medicine has even attempted. It has taught us the relatively low blood-pressure in the pulmonary artery, and has even succeeded in directly measuring this blood-pressure, with the result of proving that the relation between the pulmonary and carotid pressure varies in different species of animals, being in the rabbit $\frac{1}{4\cdot2}$, in cat $\frac{1}{5\cdot3}$, in dog $\frac{1}{3\cdot05}$.

Again, physiology has established the existence in one species, the dog, of vaso-motor nerves which exercise more or less control over the blood-flow through the pulmonary artery.

In a valuable paper communicated to the Royal Society in February, 1889, Dr. Bradford, *George Henry Lewes Student*, and Dr. Dean, arrived at the following conclusions.

"The pulmonary vessels of the dog are supplied with vaso-motor fibres leaving the cord through the roots of the uppermost dorsal nerves. No efferent vaso-motor fibres have been detected in the vagus nerve.

"The pulmonary circulation is comparatively independent of the systemic, and alterations in the blood-pressure of the latter must be of large amount to affect the pulmonary blood-pressure."

"It is probable that no rise of aortic pressure can materially influence the pulmonary blood-pressure, unless it is so great in amount or duration that the heart-muscle and valves are unable to cope with it, and so an actual regurgitation is produced."

"It is possible that the pulmonary blood-pressure can also be affected by rises of systemic pressure causing venous distension, and hence an increased supply to the right side of the heart."

"Finally, although it is undoubted from the results of this research that the mammalian pulmonary vessels receive vaso-motor nerves, yet it is probable that the vaso-motor mechanism is but poorly developed as compared with that regulating the systemic arteries."

"In this respect it may be that the pulmonary system holds an intermediate position between the systemic arteries on the one hand and the veins on the other."

The experiments on which these conclusions rest are also interesting from other points of view. I will mention one which has a close connection with my present purpose. "When the peripheral end of such a nerve as the sixth or seventh dorsal is excited a rise of pressure in both the pulmonary and aortic system is observed. The pulmonary rise, although considerable, e. g. 3 or 4 mm. Hg, is not out of proportion to the aortic rise which, with these nerves, may be as much as 30 or 40 mm. Hg. On ascending, however, very different results are obtained. Thus in one case the fifth dorsal gave an aortic rise of 10 mm. Hg only, but the pulmonary rise was 3 mm. Hg. Clearly the latter was not a passive effect of the former. In another case the fourth dorsal gave an aortic rise of 20 mm. Hg, and a pulmonary rise of 4 mm. Hg."

" Perhaps, however, the most marked and conclusive result is seen with the third dorsal nerve. This nerve frequently causes no aortic rise, and, indeed, sometimes actually a fall, *e. g.* 10 mm. Hg, but in both these cases there is a distinct pulmonary rise of 3 or 4 mm. Hg."

Thus far then we have learnt : 1. That the pulmonary circulation, like the systemic, is, to a certain extent, under the control of vaso-motor nerves.

2. That the two vaso-motor systems are closely connected with each other, inasmuch as direct irritation of the same nerve or of the same part of the spinal cord produces a simultaneous effect on both sets of vessels.

3. That this effect may be either a rise or a fall in pressure, in both, or a rise in one and a fall in the other.

The next step was to ascertain whether the effect of certain drugs is the same, for each, upon the two circuits. The facts just mentioned make it, to say the least, highly probable that this is not the case, but the question could only be answered satisfactorily by direct experiment.

Through the kindness of Dr. Bradford, of University College Hospital, and of Mr. Bokenham, assistant to my colleague Dr. Lauder Brunton in the science workroom of St. Bartholomew's Hospital, I am enabled to lay before you the following statements as to the comparative effect of certain substances upon the two circulations. Their researches were carried on entirely independently of each other and yet their results are practically identical.

In looking at the pressure tracings taken by them one cannot but be struck by the smallness of the variations in the pulmonary, as compared with those in the carotid artery: but it must be remembered that, in consequence of the lower natural pressure in the pulmonary artery alterations in pressure in either direction produce about twice the effect upon the rapidity of the lesser circulation which corresponding changes do upon the greater.

Muscarine.—In the 'British Medical Journal' for November 14th, 1874, Dr. Lauder Brunton gives the following account of the action of muscarine on the heart. Having thoroughly narcotised a rabbit with hydrate of chloral, he commenced artificial respiration, and opened the thorax. Both sides of the heart seemed to be equally filled, the veins only moderately distended, and the lungs rosy. On injecting a little muscarine into the jugular vein everything at once changed. The lungs became blanched, the left side of the heart became small, the right side swelled up, and the vena cava became greatly distended. After a short time a little atropine was injected into the jugular vein, and everything instantly returned to its normal condition. The left side of the heart regained its former size, the right side diminished, the distension of the veins disappeared, and the blanched lungs again assumed a rosy hue. Distrusting his own personal observation Dr. Brunton got two observers who knew nothing about the experiment and repeated it before them, noting down their observations, which agreed exactly with his own. It is all but impossible, I think, either to doubt the accuracy of the record of this experiment, or to attribute the phenomena observed, especially the simultaneous distension of the pulmonary artery and the blanching of the lungs, to any other cause than the action of vasomotor nerves.

Subsequent observers seem to have failed to obtain the striking effects witnessed by Dr. Lauder Brunton, but his account is substantially confirmed by Mr. Bokenham, who tells me that in his experiments[*] muscarine in a small dose caused rapid fall in the carotid pressure, with, in most cases, a rise in the pulmonary pressure. This rise, however, was of short duration, and the pressure in the pulmonary artery soon fell to normal, whilst that in the carotid more slowly rose to its original level. A large dose produced paralysis of the heart and rapid fall of pressure in both circuits. (Plate I.)

Amyl nitrite (Plate II), given by injection into the jugular vein, or by inhalation, caused a rapid fall of carotid pressure, with simultaneous marked rise of pulmonary pressure. The absolute pressure in the

---

[*] These were performed on cats, dogs, and rabbits, the method adopted being that used by Dr. J. Rose Bradford, and described by him in 'Journ. Physiol.,' vol. x, p. 153.

carotid was sometimes at this stage less than that in the pulmonary artery. Both pressures subsequently return to their original level.

Nitro-glycerine (Plate III), injected into the jugular vein of a cat, produced results very similar to those observed in the case of amyl nitrite. The effect was more prolonged, but in no instance was it possible to reduce the absolute carotid pressure below that in the pulmonary circuit. Experiments with this drug on dogs yielded the same result.

Digitalis (Plate IV).—Digitalin (Duquesnel) produced a steady rise of blood-pressure both in the carotid and pulmonary arteries, with (in small doses) marked slowing of the pulse.

Tincture of digitalis caused, when injected, a primary fall of carotid pressure, the pulmonary pressure being at first unaffected. Subsequently, a steady rise took place in both pressures.

Infusion of digitalis, in cats, produced a temporary great rise of pressure in both circuits, followed by a slight but more permanent elevation. The primary rise was in this case probably due to the fact that a large quantity of fluid had to be introduced, as the same result followed the injection of a similar quantity of water.

Strophanthus (tincture, Plate V).—In cats a small dose (mj—mij) produced a primary fall of pressure in both circuits, which very shortly, however, returned almost to normal. The heart was rendered slow and irregular by these doses.

A large dose produced a primary great rise of carotid pressure, this being accompanied by a slight fall and subsequent rise of pulmonary pressure. **The heart-beats then became** very quick and feeble, **as** shown by a steady fall of carotid pressure. This did not affect **the** pulmonary pressure, which remained **at its** original **level.**

Ergot (Plate VI).—In both cats and dogs injections of the liquid **extract** caused a primary **fall** of **the carotid** pressure, with simultaneous rise of the pulmonary curve. Ultimately a maintained rise was observed in both circuits.

Aconitine **(Plate VII).**—This drug produced a fall **both of carotid** and pulmonary pressures.

Strychnine (Plate VIII).—Small doses caused a **rise of pressure in both** circuits, **the** pulmonary rise **being especially marked. Large** doses produced **convulsions, during which** the curves both showed a great increase **of pressure,** the height varying with **the convulsions.**

Chloroform (Plate IX).—Inhalations caused simul- **taneous** fall of pressure in both circuits.

Ether (Plate X).—Inhalations caused simultaneous rise of pressure in both circuits.

Atropine sulphate (Plate XI).—Injection of a solution of **this drug caused** first a fall of both curves with slowing **of the heart.** During this time inhibition **could be produced** by stimulating the vagus. As the vagus became paralysed the pressure rose in **both circuits, and** subsequent **doses** produced no further effect.

Quebracho tincture (Plate **XII)** caused first a fall **of** carotid pressure, accompanied by a slight rise of the pulmonary pressure. Finally, pressure rose **in** both circuits, more marked in the carotid than in the pulmonary artery.

In the graphic records of these experiments there are many things of great interest, but from lack of time I have confined myself to one point only, viz. the pressure relations of the two circuits to each other. And it is clear that in some cases these relations, under the influence of the same drug, vary in a manner which could scarcely have been ascertained without the aid of direct experiment. It is difficult to arrive at any classification of them beyond this ; that

If the systemic pressure rises then the pulmonary pressure also rises.

If the systemic pressure falls then the pulmonary may either rise or fall.

And even this is only true of the primary effect of the drug, for among the later effects, *e. g.* in the case of amyl nitrite, and to a less extent in that of nitro-glycerin, the pressure in the pulmonary artery may fall, whilst that in the carotid is rising. Again, although it might be impossible from these pharmacological researches alone to prove the existence of pulmonary vaso-motor nerves, still, given the fact that such nerves do exist, and this has now been established, then some of their results seem to find their best, perhaps only explanation, on the vaso-motor nerve theory.

Of course in the production of these variations of pressure, vaso-motor influence is not the only one at work. They may be due, *e. g.* to increased force of the heart's action, or to causes as purely me-

chanical as anything can be which takes place
within a living organism as, e. g. when a block in the
aorta raises the pulmonary pressure, whilst a block
in the right side of the heart or in the lungs lowers
the systemic pressure.

Still the knowledge that certain conditions of
pressure in the pulmonary circuit may be determined by the administration of certain drugs, even
though we may be left in ignorance of the exact
mode in which this effect is brought about, may be
of great value.

But before attempting to appraise the value to
medicine of physiological (pharmacological) researches such as these, I have a mournful duty to
perform. I must say "good-bye" to my trusty
physiological guides, who have already fallen a little
behind and decline to go with me any further.

They tell me, and from the point of view of their
science it is their plain duty to do so, that we have
no right to assume that the results of experiments
on animals, performed under conditions very different from those of healthy life, would hold good in
the case of man, that no scientific proof of the
existence of pulmonary vaso-motor nerves in the
human subject has been given, and that it is very
unlikely that such proof ever will be given. Thus
it is difficult to imagine a more perfectly unscientific
proceeding than to arrive at indications for treatment by virtue of assumptions which have no scientific basis whatever. That to do so is to step from
the rock on to, and into, the quicksand. Well,
until physiology holds out to us a more hopeful pros-

pect than this, and more effectual aid than she now does, the path of our duty leads straight through the quicksand. We must make the most careful, and the best use of the data, such as they are, which we possess, and may then after all entertain good hopes that the conclusions which we may come to, although, logically, they must always remain probable only, may yet be true.

Now, personally, my evidence in the matter is prejudiced. I confess I should like to have myself a vaso-motor pulmonary arrangement with the latest improvements up to date. Such a mechanism, if it exists, must be a great comfort and advantage to its possessor. It would regulate, from the most favourable standpoint, the supply of blood to the left side of the heart meeting and satisfying the various and ever-varying needs of the organism with the right quantity of freshly aerated blood, and at the same time safe-guarding the delicate tissue of the lungs from the dangers which wait upon congestion. For when, as must often happen, the blood accumulates to more or less excess in the pulmonary circuit and in the great systemic venous trunks, then the action of pulmonary vaso-motor nerves would, *pro tanto*, tend to transfer the increase of pressure from the former to the large veins of the trunk, and to easily distensible organs such as the spleen and liver.

In support of this let me refer you again to Dr. Lauder Brunton's experiments with muscarine, which prove that, at all events in some of the lower animals, a mechanism exists which is capable of accomplishing a great deal in the direction of relieving the pulmonary veins and capillaries from undue pressure.

Of course this line of argument, a very tempting one to some people, is utterly worthless, and if no other and better proof can be adduced, the case must be given up. Such an argument, however, may be found even in the fact that pulmonary vaso-motor nerves of more or less activity have been demonstrated in the dog, i. e. certain nervous arrangements have been found to exist in that animal in connection with the function of respiration, a function necessary to life and carried on by very similar mechanism in many thousands of species.

Is it not highly probable that this same nervous mechanism exists in all, or almost all animals whose respiratory apparatus is similar and their type of development not inferior to that of the particular species in which its existence has been demonstrated ?

So far as man is concerned, the argument gains additional strength, its conclusion attains to a higher degree of probability, when we remember that it is impossible to doubt the existence in him of systemic vaso-motor nerves at least as fully organised, as richly endowed, as those which have been demonstrated in lower animals by direct experiment, and to which in man, as in the dog, a pulmonary vaso-motor system would be a fitting complement.

Surely there is uniformity in organic, not less than in inorganic nature : a uniformity which enables us to forecast with confidence the existence of similar functions and structures in animals specifically or generically allied.

I have no wish to deny that conclusions arrived at by such a process are, after all, only probable, but their probability is eminently reasonable, if due care be taken in the verification and use of the facts on which they rest : the argument must start from facts recognised as such by science and not from mere hypotheses. In this way we may hope to maintain the wholesome and necessary relations between medicine and physiology : for it will be an evil day for both, if ever the results of laboratory work are held to have no bearing upon medical practice.

The not improbable hypothesis that pulmonary vaso-motor nerves exist in man may be used to explain some otherwise puzzling facts in connection with disease of the heart, and especially of the mitral orifice. Cases are met with not infrequently in which the primary seat of the obstruction to the blood-flow is on the left side of the heart, and there is no evidence of anything wrong on the right side. At the same time the cervical veins are distended, there is a certain amount of general anasarca, and exaggerated loudness of the second sound of the heart, in the second left intercostal space, makes it probable that there is undue pressure in the pulmonary artery. And yet the lungs, in whose veins and capillaries the effects of the cardiac difficulty must, one would think, be first felt, show but slight, if any, signs of congestion.

No doubt many explanations might be given of

this state of things, but none meet the conditions of the problem more exactly than contraction of the pulmonary arterioles under some vaso-motor influence. Indeed, so far as the vessels are concerned, we may fairly compare the effect on the pulmonary artery of obstruction on the left side of the heart with that of cirrhosis of the kidneys on the systemic arteries.

There is one conclusion to be drawn from the results of the pharmacological researches just laid before you which is of no small practical interest. If it be true, to use the statement in one of the best monographs on diseases of the lung, that "it is of great importance to relieve blood-pressure in hæmoptysis," then aconite ought to be a much more efficient remedy for that affection than ergot.

On this occasion your Harveian Orator enjoys great latitude in the choice of subjects, and in his mode of treating them. I trust that in your judgment I have not abused that privilege.

The relation of physiology to medicine, at all times a matter of supreme importance to our profession, is especially so at present when from the rapid scientific development of physiology the two lines of study are year by year becoming more and more widely separated, not to say antagonistic; a state of things which is distinctly injurious to the scientific claims of medicine. The remedy is in our own hands. It is our duty now more than ever, to work out and to utilise to the utmost the medical aspects of each fresh physiological advance, and we must bring our best powers to this task.

I have ventured to bring under your notice the "Conditions of the Pulmonary Circulation" as a subject in which much yet remains to be done by physiology, and from which medicine may hope to derive no little benefit. Let me add that the results of such a research may be great or small, but until they have been realised by science, and have found their place in medicine, Harvey's labours will not have yielded up their full fruit.

NOTE.

Since the report of this Oration appeared in the journals, Dr. Gustav Hofmann, of Neudek, Bohemia, has courteously called my attention to a paper published by him in the 'Allgemeine Wiener Medizinische Zeitung,' 1888, and has underlined one paragraph which I give in the original German as well as in English, lest it may perchance have suffered at my hands in translation.

" Moreover, in face of all remedies which are said to evoke energetic contraction of vessels (which in arterial hæmorrhage would be especially indicated) there is room for a justifiable scepticism.

" For of all remedies to which is ascribed, rightly or wrongly, such vessel-contracting influence, it is certainly not proved that they also bring about contraction in the vessels of the lungs. From the fact that a certain drug narrows the uterine and intestinal vessels, must not yet be concluded that it narrows all sets of vessels. It must, on the contrary, be concluded that the narrowing of the one set of vessels goes along with dilatation of the other. Of the vessels of the lungs, especially, we have no foundation whatever for the assumption that they may be so easily variable. For what would happen were the vessels of the lungs brought so easily and so quickly to contract, and at any moment so enormous a damming in the pulmonary circuit, with dilatation of the right side of the heart, could come to pass?"

" Aber auch allen Mitteln gegenüber, welche eine energische Contraction der Gefässe hervorrufen sollen, was bei arterieller Blutung ja besonders augezeigt wäre, ist ein berechtigter Skepticismus am Platze. Denn von allen Mitteln, denen man einen solchen gefässcontrahirenden Einfluss mit Recht oder mit Unrecht zuschreibt, ist durchaus nicht erwiesen, das sie auch die Lungengefässe zur Contraction bringen. Daraus, dass ein Medicament die Uterus- und Darmgefässe verengt, darf noch nicht geschlossen werden, dass es alle Gefässpartien verenge. Es muss im Gegentheile geschlossen werden, dass die Verengerung der einen Gefässpartie mit Erweiterung der anderen einhergehe. Von den Lungengefässen im Besonderen haben wir gar keinen Grund anzunehmen, dass sie so leicht alterabel wären. Wohin kämen wir denn, wenn die Lungengefässe so leicht und so rasch zur Contraction zu bringen wären und jeden Augenblick eine so colossale Stauung im Lungenkreislaufe mit Dilatation des rechten Herzens zu Stande kommen könnte?"

PRINTED BY ADLARD AND SON, BARTHOLOMEW CLOSE.

I.

CAT 31·7·90.

MUSCARINE.

$\frac{1}{400}$ gr in femoral Vein.

Muscarine.
1/100 gr.

0"   15"   30"   45"   1'   2'   3'

Time

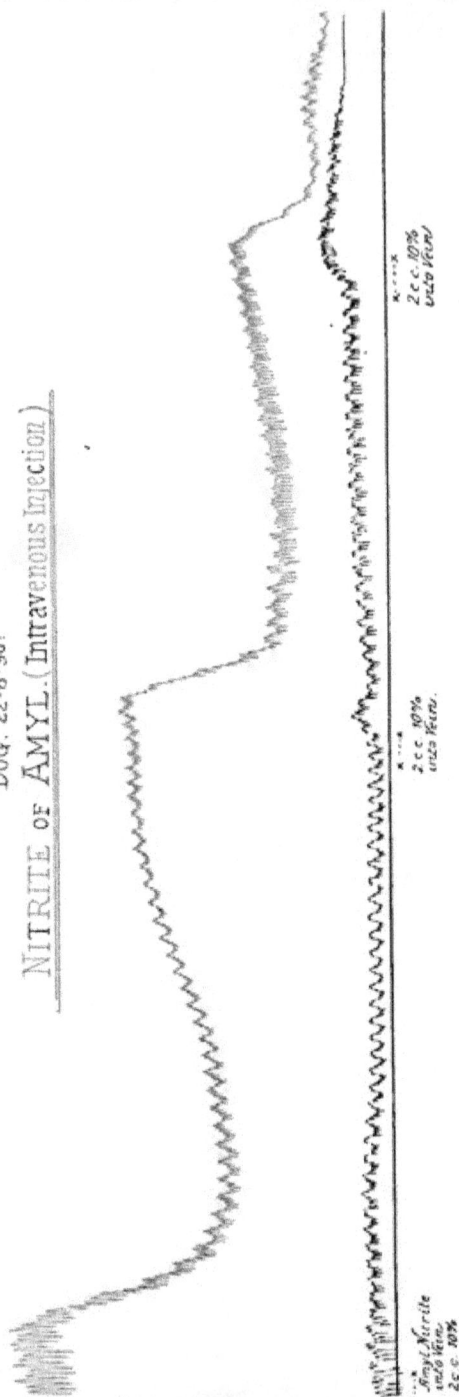

2.

Dog. 22·8·90.

NITRITE of AMYL. (Intravenous Injection.)

Amyl Nitrite
into Vein.
2 c.c. 10%

2 c.c. 10%
into Vein.

2 c.c. 10%
into Vein.

Time in Seconds

Dog - 21·8·90.
NITROGLYCERINE.

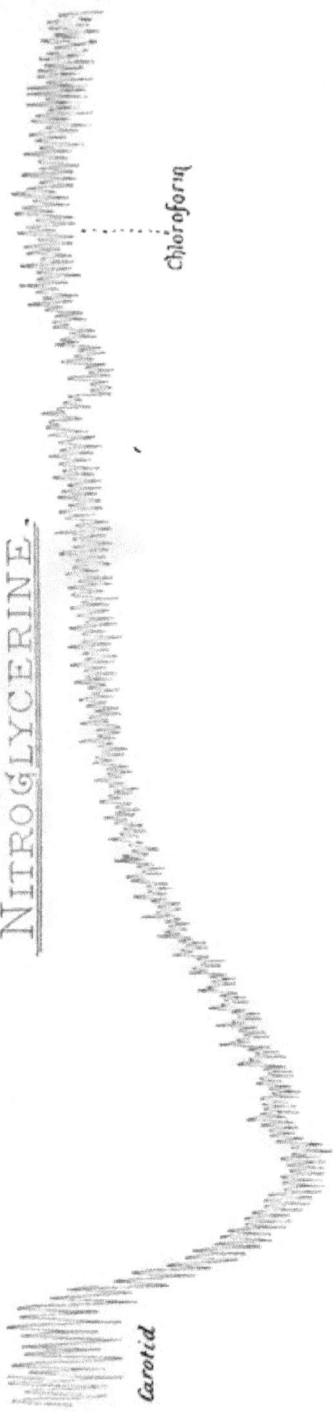

Carotid

Chloroform

Pulmonary

Abscissa
* 1 c.c. NITROGLYCERINE 1% SOLUTION

Time in Seconds

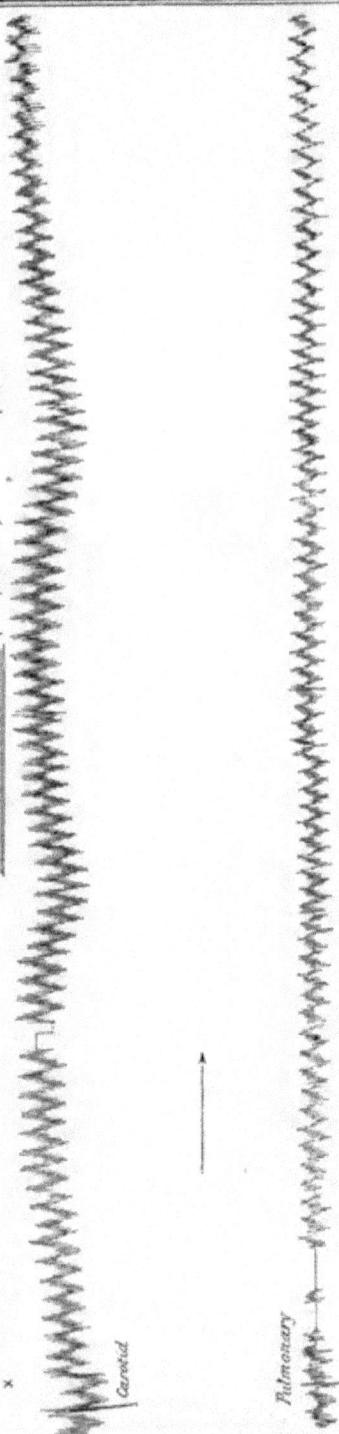

Dog. 9·7·90.

DIGITALIN (Duquesnel)

¹⁄₂₀ gr. in jugular
×

Carotid

Pulmonary

Abscissa.

Water through
+ jugular vein.

¹⁄₂₀ gr.
×

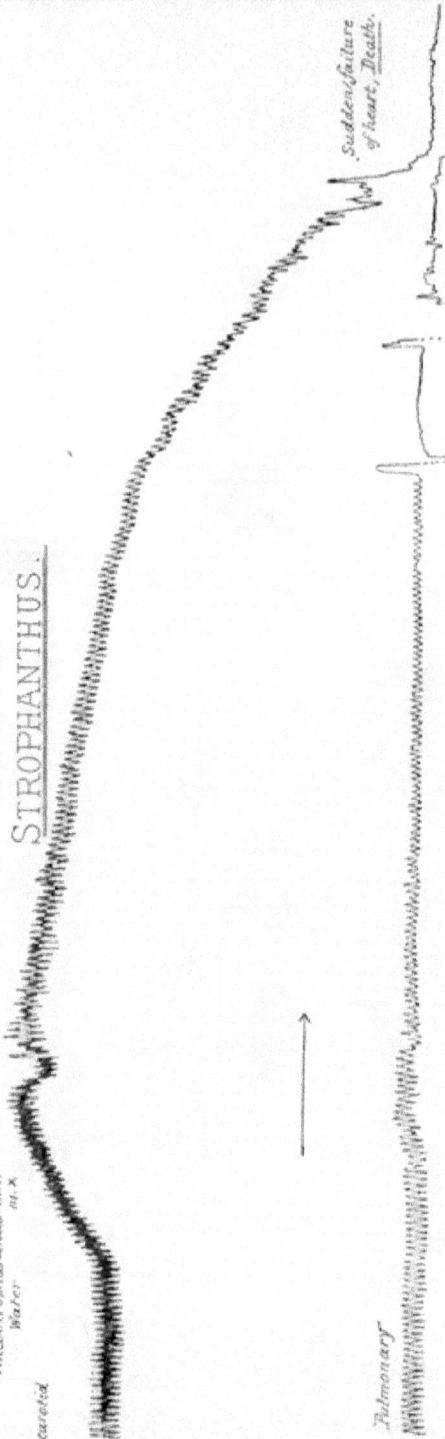

Cat. 24-7-90

STROPHANTHUS.

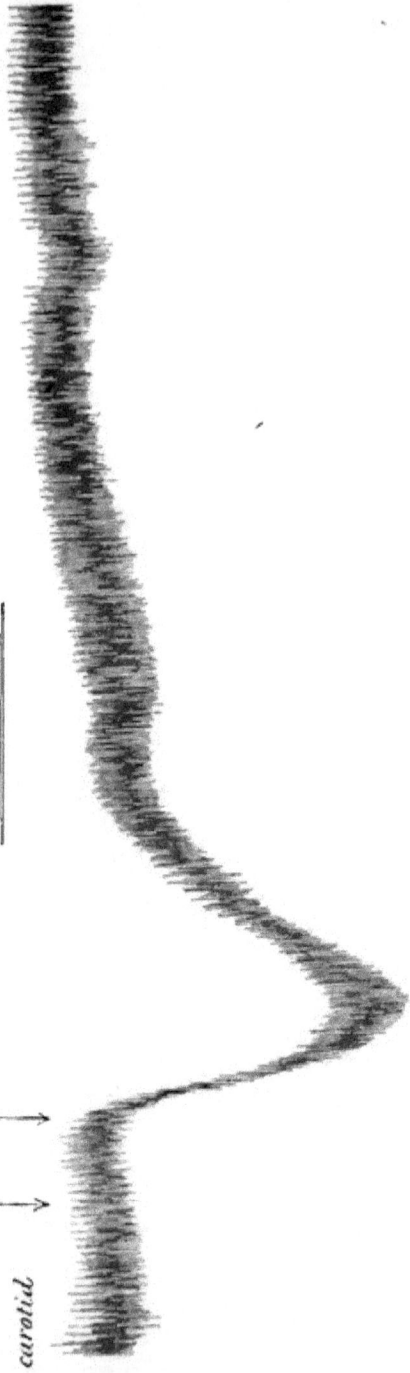

Dog. 23-7-90
ERGOT.

6.

Ext. Ergot Liq. m.X
m.XXX
Water

carotid

Pulmonary

Abscissa

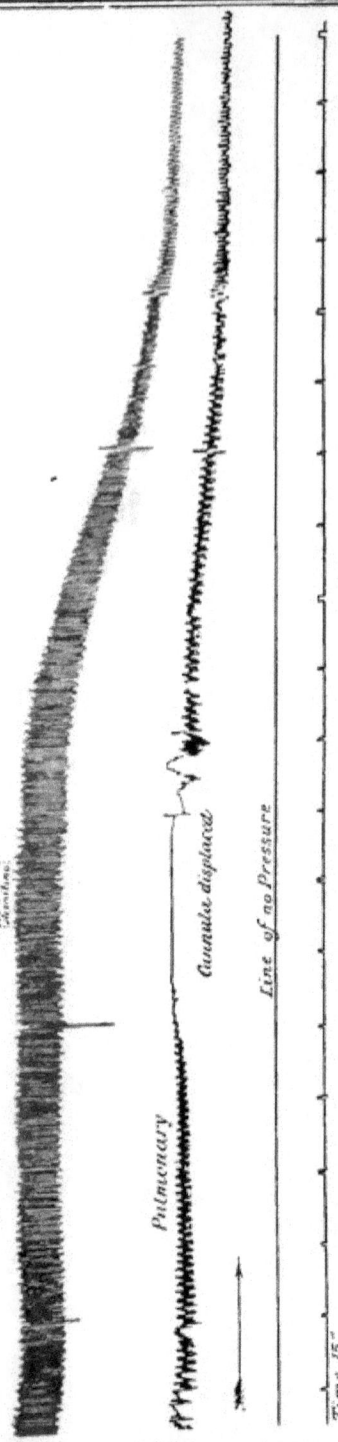

Cat. 28.7.90.
ACONITINE.

$\frac{1}{20}$ Gr.
Aconitine

7

carotid

Pulmonary

Cannula displaced

Line of no Pressure

Time 15″

8

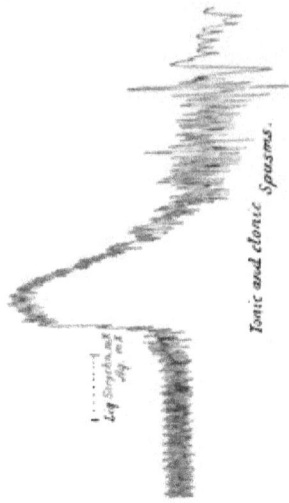

Dog. 18·7·90

STRYCHNINE

Strychnine

Liq Strychnine $\underset{2g}{m}$ X.

Tonic and clonic Spasms.

carotid

Pulmonary

Abscissa

Time 0'   30'   60'   90'

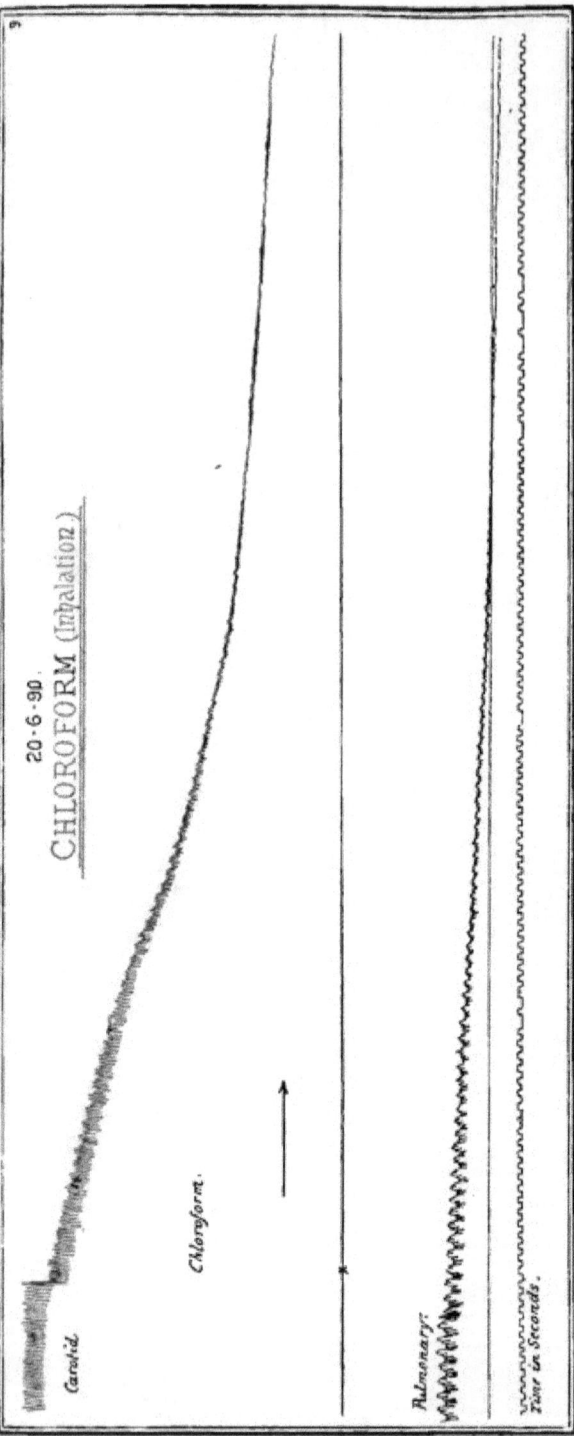

20·6·90.
CHLOROFORM (Inhalation)

Carotid.

Chloroform.

Pulmonary.

Time in Seconds.

10

20·6·90.

ETHER (Inhalation)

Ether
inhalation.

Carotid

Pulmonary

Time in Seconds.

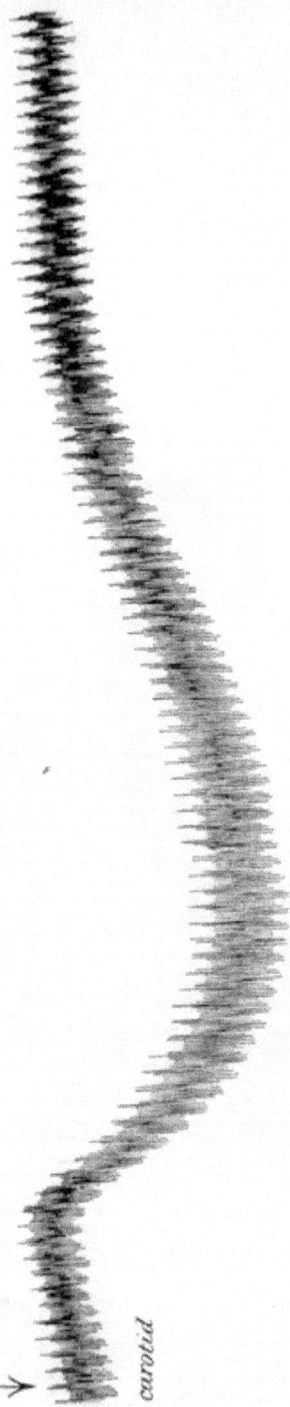

Dog 27·9·90

ÁTROPINE SULPH

ATROPINE sulph
⅒ Gr →

carotid

Pulmonary →

Abscissa

Bitch 3·10·90.

Tinct. QUEBRACHO.

Ether Bottle refilled

Temporary Cessation of Respiration

Tr. Quebracho m XV in Water m XXV.

carotid.

Pulmonary

Abscissa.